Institutional Hospital Activities for Pharmacy Practice

First Printing: 2020

ISBN: 978-1-67819-461-1

Affiliation of Author

Dr. Sagar Pamu
Assistant Professor,
Guru Nanak Institutions Technical Campus-School of Pharmacy,
Ibrahimpatnam, Ranga Reddy District, Hyderabad.

Ramya Gangannagudem
Guru Nanak Institutions Technical Campus-School of Pharmacy,
Ibrahimpatnam, Ranga Reddy District, Hyderabad.

Hemaja Anugandula
Guru Nanak Institutions Technical Campus-School of Pharmacy,
Ibrahimpatnam, Ranga Reddy District, Hyderabad.

Valentina Gogoi
Guru Nanak Institutions Technical Campus-School of Pharmacy,
Ibrahimpatnam, Ranga Reddy District, Hyderabad.

www.lulu.com
Lulu Press, Inc
627 Davis Drive, Suite 300, Morrisville, NC 27560.

Institutional Hospital Activities for Pharmacy Practice

Author

Sagar Pamu

Ramya Gangannagudem

Hemaja Anugandula

Valentina Gogoi

Editor

Sagar Pamu

2020

About the Author

Dr. **Sagar Pamu,** Assistant Professor of GNITC-School of Pharmacy. He authored 20 books, 20 research, and case report articles in national and international publications. He also filed one patent. He also participated and presented poster & oral in various national and international conferences, workshops or symposiums.

Ms. Ramya Gangannagudem, Ms. Hemaja Anugandula, and Ms. Valentina Gogoi are the undergraduates of Pharm. D at GNITC-School of Pharmacy. They have experience and skills in ADR detection, Patient Counselling, and Case Sheet Monitoring and identifying medication errors and intervention management.

Acknowledgments

At the very outset, we thank God, the Almighty for showering his blessings and being a source of guidance and wisdom throughout the study without which no human achievement is possible.

We are indebted to our beloved **Parents** without whose encouragement and help our professional career would never see the light of the day.

As we walk along the path of life, we have the pleasure of meeting people who search our life in such a way that it never is the same gain...it may be small thoughtful things they do, a smile, a helping hand, a word of encouragement or just by mere presence they make our life worth living.

Accomplishing this project has been a great learning and a very fulfilling experience. There have been many people who have come alongside and helped in conceiving, designing, and executing this project. I would like to place a record and my sincere appreciation for their contribution.

I thank our Lord God Almighty for His gracious blessings. And I dedicate this book to my beloved Parents & Siblings

Table of Contents

Introduction

Pharm. D or Doctor of Pharmacy is a professional degree in pharmacy stream. Pharm. D course duration is six years in total inclusive of 5 years of learning and one year of internship for practical learning. Pharm. D professionals need skills to provide better clinical pharmacy services such as up-to-date knowledge of clinical aspects of drugs and good communication skills.

Pharm. D undergraduates will be allotted in hospital postings in various departments designed to complement the lectures by providing practical clinical discussion; attending ward rounds; follow up the progress and changes made in drug therapy in allotted patients; case presentation upon discharge. As per their curriculum, students are required to maintain a record of cases presented and the same should be submitted at the end of the course for evaluation. A required number of cases should be presented and recorded covering the most common diseases.

For the need of documentation during their hospital postings, institutional hospital activity forms will be more helpful. So that learners will be able to develop the skills of self-learning critical thinking, problem identification, and decision making. The institutional activity forms have been designed to focus on the learning process of solving drug therapy problems, rather than simply finding the scientific answers to the problems themselves. Learners learn scientific facts during the resolution of case study problems, but they usually learn more of them from their own independent study and from discussions with their peers than they do from the instructor. The learner's job is to work through the facts of the case, analyze the available data, gather more information, develop hypotheses, consider possible solutions, arrive at the optimal solution and consider the consequences of the learner's decisions.

The role of the teacher is to serve as a coach and facilitator rather than as the source of the answer. In fact, in many cases, there is more than one acceptable answer to a given question and instructors possess the correct answer. Rather, the students become teachers and learn from each other through a thoughtful decision of the case.

The institutional hospital activity forms will be used as the focal point for independent self-learning by individual students and for in-class problem-solving discussions by student groups and instructors, if meaningful learning and discussion are to occur, students must come to discussion sessions prepared to discuss the case material rationality, to propose reasonable solutions and to defend their therapeutic plans. This requires a strong commitment to independent self-study prior to the session.

Types of institutional hospital forms with their general uses discusses the following

1. "Case Sheet" will be useful to collect the data of subjective, objective, assessment and plan and also to note the overview

information of drug interactions, adverse drug reactions, interventions, patient counseling.

2. "Drug Interactions" forms will be useful to collect the detail data of drug interactions which may be affecting the objective drug due to precipitant drug.

3. "Adverse Drug Reaction" form will be useful to collect the information of an unexpected/unwanted/dangerous reaction caused by the administration of the drug.

4. "Drug Intervention" form will be useful to note the action or process of intervening in a therapeutic situation which can alter the therapeutic outcome or cause unwanted effects.

5. "Patient Counselling" form will be useful to assess the patient's perception of their disease condition to clarify disease and to counsel and note the points regarding diet, exercise, interactions, storage and management of side effects.

6. "Drug Information and Documentation" form will be useful to request and document the doubts regarding drug information from any healthcare professional for their requested query.

7. "Patient History" form will be useful to record the patient medication history, medical history, and social history.

8. "Patient Referral Documentation" form will be useful to document and to refer the patient by the healthcare professional for their specific therapeutic problem or condition as for opinion.

9. "Patient Counselling Quality Assurance" form will be useful to assess the quality of patient counseling for the patient's condition and drug administration.

10. "Prescription Slip" form will be useful to assess and record the prescription of drugs.

Activity - 1
Patient Profile Details

The case method is used primarily to improve the skills of self-learning, critical thinking, problem identification, decision making, learning the method of solving a drug therapy problem. Case studies in the health sciences provide the personal history of an individual patient and information about health problems that must be cured and solved. The pharmacist's job is to work through the facts of the case, analyzing the available data, gathering more information, developing hypotheses, considering possible solution and arriving at the optimal solution. Patient profile form is usually written with subjective, objective, assessment and plan information (i.e. SOAP format).

Subjective Information:

Demographic Details

It includes information about a patient, consultant doctor department, age, sex, date of admission (DOA), date of discharge (DOD), height, weight, body mass index (BMI), blood group (BD).

It contains all the patient demographic details which used to be filled by the pharmacist.

Chief Complaints:

The most significant or serious symptoms (or) signs of illness (or) dysfunction that cause him or her to observe health discomforts were given in a subjective description. The chief complaints should be written by using patient words (or) statements in the patient's own words. The patient complaints should be recorded by the doctor (or) physician by listening (or) contacting directly with the patient. The chief complaints should not be considered as final diagnosis. The chief complaints should be recorded using the patient's words along with the period and replacing the patient's words with their diagnostic interpretation should be avoided.

(Example: A 60 years old male patient with decreased urine outflow, vomiting eventually was found to have bilateral facial edema)

Present Medical History:

The present medical history is complete information about patient syndrome. It includes date of onset, precise location, nature of onset, severity, and duration, presence of exacerbations and remissions, the effect of any treatment given, relationship to other symptoms bodily functions or activities (example activity, meals), degree of interference in daily activities.

(Example: bilateral facial edema sepsis with multiple organ dysfunction, B/L renal parenchymal disease oliguria)

Past Medical History:

It includes information about the patient's serious illnesses, surgical procedures, and injuries that the patient has experienced previously. Minor complaints   such as influenza, cold are usually not considered unless they show their effect on the current medical situation

(E.g. Diabetes mellitus type 2)

Past Medication History:

It includes information about the patient's past medication history and usage. It should be filled if the patient was given medication under past medical history. It is important to obtain complete information about the patient's medication history that includes all names, doses, route of administration (ROA), schedules and duration of therapy for all medication, including dietary supplements and other alternative therapies.

(Example: Clopidogrel, metformin)

Social History:

It includes information about the social characteristics of the patients, environmental factors and behaviors that contributing to the development of the disease. It consists of a patient's marital status; number of children; educational background; occupation physical activity; hobbies; dietary habits and use of tobacco, alcohol and other drugs.

Allergies: It includes allergies to drugs food, pets and environmental factors like grass, dust, pollen that should be recorded. It should also include an accurate description of the reaction that occurred.

Family History:

It includes information about Family history like age and health of parents, siblings, and children. In the case of deceased relatives, the age and cause of death should be recorded. In the case of particular, heritable diseases and those with hereditary tendency should be noted (e.g. diabetes mellitus, cardiovascular diseases, rheumatoid arthritis, malignancy, obesity)

Surgical History:

It includes information about history related to previous surgeries done to the patient related to past medical history. It should contain all the information regarding the type of surgeries the patient undergone. It helps understand the patient's health condition.

Objective Information:

Physical Examination:

Physical examination is a test done by the doctor to determine the general status of health by looking for signs of a disease or ailment.

Vital Signs:

Vital signs show how well our body is functioning and it includes (Blood pressure, pulse rate, respiratory rate, temperature, heart rate) this helps to find some of the diseases like hypertension and patient health condition

Systemic Examination:

The systemic examination includes different assessments to indicate what is going on in the body it can be filled by the pharmacist after a complete assessment of the patient by this we can conclude the condition of the patient and systemic examination includes (CVS, RS, Abdomen, GIT, GU, CNS, Clubbing, anemia, Pallor, edema)

Provisional Diagnosis:

Provisional diagnosis is the information in which the Doctor is not 100% sure of a diagnosis because he requires more information. It can be diagnosed only by the information getting from the chief complaints, vital signs and systemic examination

Lab Investigation:

It is the medical procedure which involves the testing a sample of blood, urine, or other substance from the body. It is helpful to determine the diagnosis, plan treatment, checking to see if treatment is working or monitoring the disease over time.

Biochemistry Report:

In the biochemistry report includes the values of electrolytes present the body that are sodium, potassium, chloride, calcium, FBS, RBS, serum creatinine, blood urea.

Complete Blood Picture:

A complete blood picture is a blood test useful in evaluating the overall health condition of the patient and detecting a wide range of disorders, including, anemia, infection, and leukemia. A Complete blood picture test measures several components and features of blood including Hb, PCV, RBC Count, WBC Count, Platelet, Neutrophils, Lymphocyte, Monocytes, Eosinophils, Basophils, ESR, MCV, MCH, MCHC. If these components are increased or decreased more than the normal value it may lead to serious complications. This information should be filled by the pharmacist.

Complete Urine Examination:

The complete urinalysis provides several measurements of abnormalities in the urine. Abnormal results from this test can be indicative of several conditions including kidney diseases, urinary tract infection or elevated levels of substances which are removed by the body through excretion and the complete urine examination includes urine colour, Appearance, Reaction, and Specific Gravity, Protein, sugar, Ketone bodies, Bile salts, Bile pigments, Urobilinogen, blood, Epithelial cells, Pus cells, RBC, Casts, Crystals, others.

Liver Function Test:

A liver function test is done to diagnose and monitor liver diseases or damages. The test measures the levels of certain enzymes and proteins in the blood. Other liver function test measures enzymes in which liver cells release in response to damage or disease. The pharmacist diagnoses the disease by the increased or decreased value of the components released by the liver like total bilirubin, direct bilirubin, indirect bilirubin, SGPT (ALT), SGOT (AST), total protein, albumin, globulin, A/G Ratio, Alkaline Phosphate.

Thyroid Function Test:

Thyroid function tests are a series of blood tests used to measure how well our thyroid gland is working. Available tests include the T3, T4, and TSH, If there is an increased value of the T3, T4, TSH the pharmacist will diagnose the problem of the thyroid gland and this values can be noted by the pharmacist for the further evaluation of the disease condition some of the examples of the thyroid function test.

Lipid Profile Test:

A complete cholesterol test is also called a lipid panel or lipid profile test. The doctor uses it to measure the amount of "good" and "bad" cholesterol in the body. The optimal levels of lipids show the healthy condition of the patient, increased levels of fats, triglycerides, lipids leads to certain health-related problems like obesity, heart-related problems and also increased risk factor for diabetes mellitus and hypertension.

Culture and Sensitivity:

It is a test to find germs (such as bacteria or a fungus) that cause infection. A sensitivity test is done to see what kind of medicine, such as an antibiotic, antibacterial, antifungal based on the type of infection will work best to treat the illness or infection of a particular patient. If no germs grow the culture is observed to be negative. If the germ grows the culture is positive.In terms of filling the culture and sensitivity we should mention the causative organism (germ) and about the infection of the culture and sensitivity column.

Ultrasound (U/S):

It is a medical test that uses high- frequency sound waves to capture live images from the inside of the body. It also known as sonography. An ultrasound allows the Doctor to see problems with organs, vessels, and tissues without needing to make an incision

(for example, ultrasound is done in pregnancy time to know the heartbeat of the fetus l or baby growth)

X-RAY /CT SCAN/ MRI:

X-RAY: It is a type of radiation that can pass through most solid material. It is used by the Doctor to examine the bones or organs inside the body and it is primarily written communication between the radiologist interpreting the imaging study and the physician who requested the examination. Typically, this radiology report is sent to the physician who studies the case and then conveys the result to the patient.

(CXR: R leg cellulitis)

CT SCAN: The CT scan is a combination of a series of X-Ray images taken from different angles around our body and uses computer processing to create cross-sectional images of the bone, blood vessels and soft tissues inside our body. CT scan images provide more detailed information than the X-ray

MRI: It is a test which uses powerful magnetic radio waves, and a computer to make detailed picture inside the body. The Doctor can use this test to diagnose patient or to see how well patient is responding to the treatment

ECG / 2D ECHO:

This test is done to measure the electrical activity of the heartbeat of the patient, with each beat, electrical impulses travel through the heart. These waves of ECG causes the muscles to squeeze and pump blood from the heart. The pharmacist will perform the ECG when it is necessary. If there is any evaluation or degradation of any waves it should be noted in the case sheet and maintained for better health checkup of the patient

Assessment Information:

Assessment

Assessment of the patient condition should be done by the subjective, objective and chief complaints, the focused physical exam should include the following components: (Test results, assessment of physical, mental, neurological status and vital signs, airway assessment, lung assessment, CNS and PNS assessment). Assessment affects treatment by narrowing down diagnostic impression effectiveness. The Assessment of the case affects outcomes by ensuring that patients are transported to proper receiving facilities.

For example, The patient was admitted to the hospital in the case of sepsis, oliguria, on X-Ray he was observed to have cellulitis on the right leg and was prescribed with anti-infective drugs and certain suitable medication.

(Example: The patient was admitted to the hospital in the case of sepsis, oliguria. On X-Ray he was observed to have cellulitis on R leg and was prescribed with anti-infective drugs and certain suitable medications)

Final Diagnosis:

The final diagnosis is confirmed after getting the results of the test, such as blood tests and biopsies that are done to find out certain diseases, infections and abnormal conditions of the patient.

Plan Information:

Physician's Medication Chart:

The medication chart prescribed by the physician

Trainee Pharmacist Medication Chart:

It is a medication chart that is filled by the trainee pharmacist which should be discussed with the doctor or physician and then it should be dispensed.

Example:

S.No	Drug Name	Generic Name	ROA	Dose & Frequency	Therapeutic Use	Duration Date
1	Inj. cenems	Cefotaxime	IV	1.5 gm I-0-0	Antibiotic / antibacterial	√
2	Inj. Thiamin	Thiamin	IV	1amp I-0-I	Treat lack of timing	√
3	Inj. Perinorm	Metoclopramide	IV	1MPR I-0-0	Antiemetic	√
4	IVF-NS, RL	Sodium chloride	IV	50ml I-0-0	Electrolyte supplement	√
5	T. Nodosis	Sodium bicarbonate	P/O	500mg I-0-0	Treat heartburn	√
6	T. Ketoalpha	Keto analogues	P/O	1 mg I-I-0	Treat kidney diseases	√
7	Inj. PAN	Pantoprazole	IV	40mg 0-I-0	Proton pump inhibitor	√
8	Inj Dalacin	Clindamycin	IV	600 mg I-I-I	Antibiotic	√
9	Inj. Claclexane	Enoxaparin	IV	40mg 0-I-0	Anticoagulant	√
10	T. Chymoral -forte	Chymotrypsin	P/O	500 mg I-I-I	Reduce redness, swelling	√

11	Syp. Kcit	Potassium citrate +citric acid	P/O	2+8 p I-I-I	Treat acidosis	√
12	IVF-RL+KCL	Sodium chloride + potassium chloride	IV	5Oml 0-I-0	Electrolyte supplement	√
13	T. Cinod	clinidipine	P/O	5mg 0-I-0	Control BP	√

Drug-Drug Interaction Screening:

It includes all information about drug interactions between two drugs, their severity, effect, and drug management. It should be filled after observing any side effects and adverse effects. Sometimes it can also be filled in case of familiar drug interactions.

Adverse Drug Reactions:

It includes information about the adverse reactions caused by the drug. It should be filled carefully and completely by physicians with complete information.

Trainee Pharmacist Intervention:

IT includes information based upon the medication chart given by the physician. It should be filled when the medication chart contains incomplete RX or additional medication, replacement of drugs, duplicate drugs.

Discharge medication:

It includes information about the medication given to the patient in discharge time. It should be filled by the doctor or physician or doctor. Discharge medication is mostly composed of drugs that are taken in oral form like syrups, etc.

For example:

S.No	Drug Name	Generic Name	Dose	ROA	Frequency	Indication	Duration
1	T. Chymoral forte	Chymotrypsin	500mg	p/o	TID	Reduce redness and swelling	I-I-I
2	T. Nodisis	Sodium bicarbonate	50mg	p/o	OD	Treat heartburn	I-0-0
3	T. Keto alpha	Keto analogs	1mg	p/o	BID	Antibacterial	I-I-0
4	T. cinod	Clinidipine	5mg	p/o	OD	Control BP	0-I-0

Discharge Advice:

It includes the information about the discharge medication given to the patient during discharge mostly it includes oral drugs or syrups for easy convenience of the patients for taking medications

Patient Counselling:

It is based upon the diet and lifestyle modification of the patient. it should be explained by the physician or pharmacist to the patient. Patient counselling also includes precautions and preventions. it should be clearly explained to the patient for better health conditions.

(For Example, Alcohol should be avoided normal and soft diet should be prescribed to the patients, low lipid diet should be taken)

Signature of Trainee Pharmacist:

It should be signed by the trainee pharmacist.

Signature of staff:

It should be signed by the staff member or senior pharmacist.

Remarks:

In case of any remarks found by the staff or senior pharmacist. It should be explained clearly under this remarks column.

References:

1. Herreid CF. case studies science: a novel method of science education. J Coll Sci Teach 1994; 23:221-229.

2. Pharmacotherapy case book, a patient-focused approach 8th edition by Terry L. Schwinghammer, Julia M. Koehler.

Patient Profile Form

Subjective Information

Demographic Details

Case No:

Patient Name:		IP/OP No.:	
Sex:		Age:	
DOA:		DOD:	
Consultant:		Department:	
Ht:	Wt:	BMI:	BG:

Chief Complaints:
Present Medical History

Past Medical History:	Past Medication History:

Social History:

Smoking : Yes/No; if yes _____ packs/day
Alcohol: Yes/No; if yes _____ Quantity
Chewing tobacco: Yes/No; if yes _____ Quantity

Allergies:

Food: (Veg/N.Veg)
Drug:
Others:

Family History:	Surgical History:

Objective Information

Physical Examination

Vital Signs:

Date							
Blood pressure (mmHg)							
Pulse Rate (/min)							
Respiratory Rate (/min)							
Temperature (^{0}F)							
Heart rate (bpm)							

Systemic Examination:

CVS	
RS	
Abdomen	
GIT	
GU	
CNS	
Clubbing	
Anaemia - Pallor - Oedema -	

Provisional Diagnosis:

Lab Investigations

Biochemistry Report

Date							
Sodium: 135-146meq/l							
Potassium: 3.5-5.1meq/l							
Chloride: 95-105meq/l							
Calcium: 8.4-10.2mg/dl							
FBS: 60-110mg/dl							
RBS: <160mg/dl							
Se.Creatinine: 0.5-1.5 mg/dl							
Blood Urea: 10-50 mg/dl							

Complete Blood Picture

Date							
Hb : 11-16.5 g/dL-F 14.3-18 g/dL -M							
Pcv: 35-50 g%							
RBC count: 3.8-4.8ml/ m^3- F 4.5-6.5ml/m^3 -M							
WBC count: 4000-11000 cells/ m^3							
Platelet: 1.5-4.0 Lac/ m^3							
Neutrophils: 40-75%							
Lymphocytes: 20-40%							
Monocytes: 2-10%							
Eosinophils: 1-6%							
Basophils: 0-1%							
ESR: <20mm 1st Hr							
MCV: 80-100Fl							
MCH: 26.5-33.5							
MCHC: 31.5-35.0							

Complete Urine Examination

Date								
Colour								
Appearance								
Reaction								
Spf.Gravity								
Protein								
Sugar								
Ketone bodies								
Bile salts								
Bile pigments								
Urobilinogen								
Blood								
Epithelial cells								
Pus cells								
RBC								
Casts								
Crystals								
Others								

Thyroid Function Test

Date							
T3: 0.8-2.0ng/ml							
T4: 5.13-14.06ug/dl							
TSH: 0.46-4.7Iu/ml							

Patient Profile Form

Liver Function Tests

Date						
Tot. bilirubin: 0.22-1.0mg/dl						
Direct bilirubin: 0-02mg/dl						
Indirect bilirubin: T.B-D.B						
SGPT(ALT): Upto45U/L						
SGOT(AST): 5-40U/L						
Tot.protein: 5.5-8.0gm/dl						
Albumin: 3.5-5.0gm/dl						
Globulin: 2.0-3.5gm/dl						
A/G ratio:						
Alk. Phosphate: 30-120 U/L						

Lipid Profile

Date						
Cholesterol: <200mg/dl						
HDL: >50mg/dl						
LDL: <100mg/dl						
Triglycerides: <150 mg/dl						

Others

Culture and Sensitivity:	U/S:
X-Ray / CT Scan / MRI:	ECG / 2D ECHO:

Assessment Information

Final Diagnosis

Plan Information

Day Wise Assessment

Day 1:	Medication Chart:
B.P:	
Pulse:	
C/O:	
O/E:	
Adv:	
Day 2:	Medication Chart:
B.P:	
Pulse:	
C/O:	
O/E:	
Adv:	

Day 3:	Medication Chart:
B.P:	
Pulse:	
C/O:	
O/E:	
Adv:	
Day4 :	Medication Chart:
B.P:	
Pulse:	
C/O:	
O/E:	
Adv:	
Day5 :	Medication Chart:
B.P:	
Pulse:	
C/O:	
O/E:	
Adv:	

Day6 :	Medication Chart:
B.P:	
Pulse:	
C/O:	
O/E:	
Adv:	
Day7 :	Medication Chart:
B.P:	
Pulse:	
C/O:	
O/E:	
Adv:	
Day7 :	Medication Chart:
B.P:	
Pulse:	
C/O:	
O/E:	
Adv:	

Physician's Medication Chart

S. No.	Drug Name	Generic name	ROA	Dose & frequency	Therapeutic use	Duration							

Trainee Pharmacist Medication Chart

(A Practice Medication Chart-Not to be Dispensed)

S. No.	Drug Name	Generic name	ROA	Dose & frequency	Therapeutic use	Duration							

Drug-Drug Interaction Screening:

S. No.	Drug1	Drug2	Severity	Effect	Management

ADR:

S.No.	Drug	Reaction

Trainee Pharmacist Intervention:

S.No.	Intervention Made

Discharge Medication:

Brand Name	Generic Name	Dose	ROA	Frequency	Indication	Duration of Use

Discharge Advice:

Patient Counselling:

Signature of Trainee Pharmacist:	Signature of Staff:
	Remarks:

Patient Profile Form

Institutional Hospital Activities for Pharmacy Practice

Activity - 2
Drug Interaction

A drug-drug interaction (DDI) is precisely defined as the change in efficacy or possible toxicity of one drug by prior or concomitant administration of a second drug.

Drug-drug interaction form is designed to detect the influence of a second effective drug on the metabolism of the study drug and, conversely, the influence of the study drug on the metabolism of a second drug. DDI form is also carefully designed to identify the influence of a second drug on the exposure of the study drug and conversely, the influence of the study drug on the exposure of a second drug. The term "exposure" typically refers to key parameters of a drug's concentration within the bloodstream like AUC, Cmax, Cmin, and Tmax. The choice of the second drug is typically supported one among these specific criteria:

1. *Both specific drugs for treating the same disease.* The second drug is coadministered with the study drug, where both drugs are for managing the same disease. This in common is the possible situation with the DDI study between regorafenib (study drug) and irinotecan. Irinotecan is commonly coadministered with regorafenib where each of these drugs is for treating colorectal cancer [1].

2. *Second drug for treating a different condition.* The second drug is administered in the same timeline as the study drug, as is the case where the second drug is a contraceptive, statin, acetaminophen,

warfarin, or a monoamine oxidase inhibitor [2, 3]. In this situation, the condition to be treated by the second drug is not the same as the condition being treated by the study drug.

3. *The second drug conventionally used for PK studies.* This concerns clinical pharmacokinetic studies that are not intended to measure efficacy or safety but instead are used as a basis for predicting which class of drugs will likely engage in DDIs with the study drug. Here, the second drug is an established inhibitor of one of the cytochrome P450 enzymes and is conventionally used as a model "second drug" in drug-drug inhibition studies.

It includes all information about drug interactions between two drugs, their severity, and effect and drug management. It should be filled after observing any side effects and adverse effects. Sometimes it can also be filled in case of familiar drug interactions.

Drug Interaction Form (General Information)

It includes all the information regarding drug interactions. It includes basic demographic details like patient name, age, I.P /OP NO, DOA, weight, sex, final diagnosis

For example:
Patient name: Xyz
I.P /OP No: xxx
DOA: 11-11-19
Age: 45
Weight: 60kg
Sex: F

Final Diagnosis: Giant Lipoma

Interacting Drugs	Dose	Route	Frequency
Tramadol	50mg	IV	BID
Ondansetron	1amp	IV	BID

Objective drug:

Objective drug is the drug that is dominant or effective over the other drug.

For example, Ondansetron is the objective drug over tramadol.

Precipitant drug:

The precipitant drug is the drug that is over dominated and affected by the action of other drugs. For example, Tramadol is the precipitant drug.

Type of Interactions:

It includes the type of interactions that the drugs undergo like drug-drug interaction (or) drug-food interaction (or) chemical-drug interaction (or) drug-laboratory test interaction (or) drug-herbal interactions (or) pharmacogenetic interaction. For example Drug-drug interaction

Classification of Drug Interaction:

Varieties of mechanisms are responsible for drug interactions. They can be classified according to whether the mechanism is pharmacokinetic or pharmacodynamic. It includes two types of classification like pharmacokinetic drug interactions and pharmacodynamic drug interactions, pharmaceutical interactions. It includes the effect of drug-drug interactions through absorption, distribution, metabolism, and elimination. The effect of drug can be summoned either by additive effect (or) antagonistic effect (or) synergistic effect. It can also be chemical (or) incompatibility type of pharmaceutical interactions. For example Type of interactions

Pharmacokinetic drug interactions:

It includes information like the onset of action either rapid (or) delayed. It should also include the severity of interaction like a major (or) moderate (Or) minor. It should also be provided with the documentation like probable (or) suspected (or) possible (or) unlikely.

For example;

Onset of action: Rapid

Severity: Major

Documentation: Possible

Mechanism of interaction

The mechanism of action of two drugs should be mentioned the mechanism involved in the interaction between two drugs should be mentioned under this column

For example: On the interaction between ondansetron and tramadol, it leads to the increased risk of serotonin syndrome

Management of interaction:

The management which has to be done to avoid the consequences and adverse effects of interaction should be mentioned under this column. The management should always be in a way that it decreases the effects caused by the interaction.

For example, The patient should be observed during the treatment and administration of the drug. The drug administration should be monitored carefully in case of administration of ondansetron and tramadol due to the chances of symptoms occur within several hours

Notified to and action taken:

In case of identification of the interaction between the two drugs, it should be notified to the doctor/physician (or) action should be taken to avoid the effects caused by the interaction. If it is notified to and action is taken it should be mentioned as" Yes" if not it should be mentioned as" No".

For example:

Notified to and action taken: Yes

Signature of the pharmacist:

The signature of the trainee pharmacist should be done.

Signature of staff:

The signature of the staff should be done.

References

1. Regorafenib (metastatic colorectal cancer) NDA 203-085. Page 31 of the 64-page Clinical Pharmacology Review.
2. Package label. ADLYXIN (lixisenatide) injection, for subcutaneous use; July 2016 (33 pp.)
3. Package label. EXALGO (hydromorphone hydrochloride) extended-release tablets; March 2010 (30 pp.).

Drug Interaction Form

Patient Name: IP/OP No:

DOA:

Age: Weight:

Sex:

Final Diagnosis:

Interacting Drugs	Dose	Route	Frequency

Objective Drug:

Precipitant Drug:

Type of Interactions:

☐ Drug-Drug Interaction   ☐ Drug Food Interaction

☐ Drug Chemical Interaction   ☐ Drug Laboratory Interaction

☐ Drug Herbal Interaction   ☐ Drug Genetic Interaction

Classification of Drug Interaction:

Pharmacokinetic Drug Interaction:

☐ Absorption ☐ Distribution ☐ Metabolism ☐ Excretion

Pharmacodynamic Drug Interaction:

☐ Additive effect   ☐ Antagonistic effect   ☐ Synergistic effect

Pharmaceutical Interaction:

☐ Chemical   ☐ Incompatibility

Pharmacokinetic Interactions:

Onset of Action: ☐ Rapid   ☐ Delayed

Severity: ☐ Major   ☐ Moderate   ☐ Minor

Documentation: ☐ Probable   ☐ Possible

              ☐ Suspected ☐ Unlikely

Mechanism of Interaction:

Management of Interaction:

Notified to and Action Taken: ☐ No   ☐ Yes

Signature of Pharmacist   **Signature of Staff**

Activity - 3
Adverse Drug Reaction

An **adverse drug reaction (ADR)** is an unwanted/dangerous effect caused by taking medication [1]. ADRs may occur with a single dose or result with a prolonged administration of a single drug combination of two or more drugs. This expression differs from the meaning of "side effect", as this last expression might also imply that the effects can be beneficial [2]. The field which concerns about study of ADRs meant to known as pharmacovigilance. An **adverse drug event (ADE)** refers to any injury occurring at the time a drug is used, whether or not it is identified as a cause of the injury [1]. An ADR is a special type of ADE in which a causative relationship can be shown. ADRs are only one type of medication-related harm, as harm can also be caused by omitting to take indicated medications [3].

ADRs may be classified by e.g. cause and severity.

Cause

Type A: Augmented pharmacologic effects - dose-dependent and predictable

Type A reactions, which constitute approximately 80% of adverse drug reactions, are usually a consequence of the drug's primary pharmacological effect (e.g. bleeding when using the anticoagulant warfarin) or a low therapeutic index of the drug (e.g. nausea from digoxin), and they are therefore predictable. They are dose-related and usually mild, although they may be serious or even fatal (e.g.

intracranial bleeding from warfarin). Such reactions are usually due to inappropriate dosage, especially when drug elimination is impaired. The term 'side effects' is often applied to minor type A reactions [4].

Type B: Idiosyncratic

Types A and B were proposed in the 1970s [5] and the other types were proposed subsequently when the first two proved insufficient to classify ADRs [].

Seriousness

The U.S Food and Drug Administration defines a serious adverse event as one when the patient outcome is one of the following [7]:

- Death
- Life-threatening
- Hospitalization (initial or prolonged)
- Disability - significant, persistent, or permanent change, impairment, damage or disruption in the patient's body function/structure, physical activities or quality of life.
- Congenital abnormality
- Requires intervention to prevent permanent impairment or damage

Severity is a point on an arbitrary scale of intensity of the adverse event in question. The terms "severe" and "serious", when applied to adverse events, are technically very different. They are easily confused but cannot be used interchangeably, requiring care in usage.

A headache is severe if it causes intense pain. There are scales like "visual analog scale" that help clinicians assess the severity. On the other hand, a headache is not usually serious (but maybe in case of subarachnoid hemorrhage, subdural bleed, even a migraine may temporally fit criteria), unless it also satisfies the criteria for seriousness listed above.

Suspected Adverse Drug Reaction Reporting Form

This form gives the information about adverse drug reaction caused by the drugs.

Patient Information:

It includes basic demographic details of patient information like patient initials, age at time of event (or) date of birth, sex, weight in Kg's.

Suspected Adverse Drug Reaction:

It includes suspected adverse reactions like date of reaction started, date of recovery, describing reaction (or) problem.

Suspected medication(s):

It includes information about the suspected drug with details like brand name manufacturer (if known), batch number, expiry date, the dose used, rout used, frequency, date started, date stopped, the reason for prescribing.

1. The reaction observed after the stoppage of the drug or reduction of the dose of the drug should be mentioned. It should also include the reaction reappeared after the reintroduction of the drug.

S. No	1.Name (Brand/ Generic)	Manufactur er (if known)	Batc h No	Exp Dat e	Dose used	Rout e used	Freq uenc y	Date starte d	Date stoppe d	Reason for Prescribi ng

S. No	2. Reaction abated after drug stopped or the dose reduced					3. Reaction reappeared after reintroduction				
(C)	Yes	No	Unkno wn	Reduce d dose	NA	Yes	No	Unk now n	If Reintroduced	NA

2. Concomitant medical product including self-medication and herbal remedies with therapy dates

3. **Relevant tests/laboratory data with dates:**

Any relevant tests done (or) laboratory data collected regarding adverse drug reactions should be explained clearly under this column

4. **Relevant medical/medication history (example: allergies, race, pregnancy, smoking, alcohol use, hepatic/renal dysfunction, etc.):**

Any information regarding the above categories like Relevant medical/medication history (example: allergies, race, pregnancy, smoking, alcohol use, hepatic/renal dysfunction, etc.) should be mentioned in this column.

Reporter:

It includes the information like name of the reporter causality assessment and date of this report.

References

1. "Guideline for Good Clinical Practice". International Conference on Harmonisation of Technical Requirements for Registration of Pharmaceuticals for Human Use. 10 June 1996. p. 2.

2. Nebeker JR, Barach P, Samore MH. "Clarifying adverse drug events: a clinician's guide to terminology, documentation, and

reporting". Annals of Internal Medicine. 2004. 140 (10): 795–801. DOI:10.7326/0003-4819-140-10-200405180-00017

3. Bose KS, Sarma RH. "Delineation of the intimate details of the backbone conformation of pyridine nucleotide coenzymes in aqueous solution". Biochemical and Biophysical Research Communications. 1975.66 (4):1173–9. DOI:10.1016/0006-291X(75)90482-9.

4. Ritter, J M. A Textbook of Clinical Pharmacology and Therapeutics. Great Britain. p. 62. 2008. ISBN 978-0-340-90046-8.

5. Rawlins MD, Thompson JW. Pathogenesis of adverse drug reactions. In: Davies DM, ed. Textbook of adverse drug reactions. Oxford: Oxford University Press, 1977:10.

Suspected Adverse Drug Reaction Reporting Form

A. Patient Information

1. Patient Initials _____

2. Age at time of Event or Date of Birth _____

3. Sex _____

4. Weight in Kgs: _____

B. Suspected Adverse Reaction

1. Date of Reaction Started:

2. Date of Recovery:

3. Describe reaction or problem:

C. Suspected Medication(s)

S. No	1. Name(Brand/ Generic)	Manufacturer (if known)	Batch No	Exp. Date	Dose used	Route used	Frequency	Date started	Date stopped	Reason for Prescribing

Suspected Adverse Drug Reaction Reporting Form

S.No (C)	2. Reaction abated after drug stopped or the dose reduced					3. Reaction reappeared after reintroduction				
	Yes	No	Unknown	NA	Reduced Dose	Yes	No	Unknown	NA	If Reintroduced

4. Concomitant medical product including self-medication and herbal remedies with therapy dates (exclude those used to treat reaction)

D. Reporter:

Name:

Casualty Assessment:

Date of this report:

Activity - 4
Pharmacist Intervention

The role of pharmacists has been clinically proven to improve many outcomes regarding patient health, including greater patient safety, improved disease, and drug therapy management, effective healthcare spending, improved adherence, and improved quality of life [1]. The World Health Organization (WHO) has developed a conceptual framework to build a standard taxonomy for the patient. The WHO's new conceptual definition of patient safety reads as "the reduction of risk of unnecessary harm associated with healthcare to an acceptable minimum. An acceptable minimum refers to the collective notions of given current knowledge, resources available and the context in which care was delivered weighed against the risk of non-treatment or other treatment" [2].

The word "error", itself, brings about actions for prevention and distracts from the main goal of getting the right drug, with the right dose, with the right route, at the right time, to the right patient, a phrase known as "the five rights" [3]. Acknowledging that errors will happen due to the human condition is one thing, and then blame is easily placed on the individual, however, many experts in patient safety see new and existing errors as a fault with the systems that are in place, the systems approach assumes that several errors are inevitable and that the work environment can lead to the likelihood of certain errors occurring. However, the systems approach says nothing of an individual's responsibility to prevent medical errors and should not be seen as an excuse for a culture that relies on others to identify and resolve errors or where errors are seen as being inevitable [4].

Pharmacist's Intervention Documentation form

It is the pharmacist's documentation form which includes interventions of the drug **in** the case sheet form

1. Patient details:

It includes basic demographic details of the patients like name, date of admission, age, sex, ward, D.I. No., reason for admission, on examination, diagnosis.

For example:
Name: XYZ
Age: 45
Sex: F
Ward: cubicles
Date of admission: 11-11-19
Reason for date: Swelling over left arm, discomfort
On examination: Gaint lipoma
Diagnosis: Gaint lipoma over the left arm

2. Prescription details:

It includes all the drugs prescribed under medication charts by the pharmacists (or) doctors.

For example:

Prescription Details			3. Laboratory Data
S.No	Drug	Dose & Frequency	(This column should be filled if any intervention found under abnormal laboratory values)
1	Cefower ertapenem	1 gm BID	
2	Tramadol	50gm BID	
3	Ondansetron	1AMP BID	
4	Ultram priomepra	40mg OID	
5	IVF-NS/RL	100ml/hr	

4. Prescription problem (check all that apply):

It includes prescription problems that are found under medication charts like allergy, interaction, incomplete RX, high dose, prior ADR, unnecessary drug, duplication, low dose.

For example: Under medication chart, unnecessary drug is prescribed.

Drugs Involved	Strength	Direction	Quantity	Cost(Rs/-)
Priomepra	40mg	P/O	OID	15 Rs/-

The above table should be filled only if any intervention drug is found under the medication chart.

5. Define Prescription Problem:

This prescription problem defines the actual problem identified in the medication chart the actual reason under intervention should be explained.

For example, an extra drug is prescribed in the medication chart.

6. Action taken (check all that apply)

It is the action to be taken by the pharmacist, where the pharmacist discusses with the patient, prescriber, patient representative, drug information reference consultation and others and also to control (or) solve the intervention found.

They are like,

Discussion with patient

Discussion with prescriber

Discussion with patient representative

Drug information Reference consulted

Others (please specify)

7. Recommendations (check all that apply):

It includes the information regarding any recommendation given by the doctor (or) physician /pharmacist like changing of drug(name of the drug), dose (either increase or decrease), duration, form/Route, stopping of drug or holding of drug, the addition of drug, schedule, and others(laboratory data).

Brief:

The prescription problem should be explained briefly by the method of solving the intervention

For example, the additional drug prescribed under the medication chart should be stopped.

8. Drug Intervention Based on (Check all that apply):

It includes the information based on the drug intervention it is depending upon medication chart that includes like (no response to the treatment, inappropriate drug regimen, ADR, literature review based, laboratory data findings, and drug interaction)

For example, an inappropriate drug regimen (priomepra) is given in the medication chart.

9. Intervention Accepted:

This includes whether the intervention is accepted by the pharmacist or not.

10. If no to 8, Reason:

If there is no drug intervention in the medication chart, it implies that there is no problem in the medication chart which is prescribed by the pharmacist/doctor to the patient.

11. Attending Physician's Name & Signature:

It includes the physician's name and the signature of the physician.

12. Results (check all that apply):

Final result is included in this that is:

Rx: - dispensed as written

    Clarified & dispensed

    Not dispensed

    Changed and dispensed

For example: changed and dispensed

Patient: - counseled

> Written information is given to the patient

Other: - improved complaint

> Increased therapeutic effectiveness

> Improved monitoring of therapy

> Prevent toxicity/side effect

> Reduced cost by RS/-

By the above information states the final result of the intervention

For example, RX is changed and dispensed to the patient and written information is given to the patient as a patient counseling

13. Follow up Details of the Patient:

After the intervention, the pharmacist should follow up on the details of the patient if there are any side effects or adverse effects the medication should be stopped / replaced.

Prescription Details			3.Laboratory Data
S. No	Drug	Dose & Frequency	
1	Cefower ertapenem	1 gm BID	(This column should be filled if any intervention found under abnormal laboratory values)
2	Tramadol	50gm BID	
3	Ondansetron	1AMP BID	
4	Ultram priomepra	40mg OID	
5	IVF-NS/RL	100ml/hr	

14. Intervention Made by:

It includes the name of the person (doctor/physician) who identified and rectified the intervention found in the medication chart.

Name:
Name of the pharmacist

Designation:
Designation of the pharmacist

Signature with date:
Signature of the pharmacist with date.

References

1. Canadian pharmacists Association. 2008 http://www.pharmacists.ca/content/consumer_patient/resource_centr e/working/pdf/expanding_the_role_of_pharmacists.pdf

2. World Health Organization. 2009. The conceptual framework for the international classification for patient safety. Final technical report and technical annexures. World Health Organization Global Press.

3. Benjamin. D Reducing medication errors and increasing [atient safety: case studies in clinical pharmacology. Journal of Clinical Pharmacology. 2003; 43(7)

4. Beso AFB. The frequency and potential causes of dispensing errors in a hospital pharmacy. Pharmacy World & Science.

Pharmacist Intervention Documentation Form

1. Patient Details   D.I.No

Name: Age:

Sex: Ward:

Date of admission:

Reason for admission:

On Examination:

Diagnosis:

2. Prescription Details		3. Laboratory Data
Drug	**Dose & Frequency**	

4. Prescription Problem (Check all that apply):

☐ Allergy   ☐ Interaction   ☐ Incomplete Rx

☐ Prior ADR   ☐ Unnecessary Drug   ☐ Duplication

☐ High Dose   ☐ Low Dose   ☐ Contraindication

☐ Wrong Drug   ☐ Excessive Duration   ☐ Inconvenient

☐ Others (Please Specify):

Drugs Involved	Strength	Direction	Quantity	Cost (Rs)

5. Define Prescription Problem

----

6. Action Taken (Check all that apply):

Discussion with Patient

Discussion with Prescriber

Discussion with Patient Representative

Drug Information Reference Consulted

Others (Please Specify):

7. Recommendations (Check all that apply):

Change ☐ Drug   ☐ Dose   ☐ Duration

         ☐ Form/Route ☐ Schedule

Dose ☐ Increase   ☐ Decrease

Drug ☐ Stop/Hold ☐ Add

Others ☐ Laboratory Data

Brief:

_____

_____

_____

8. Drug Interactions Based on (check all the apply)

☐ No response to treatment ☐ Inappropriate to Drug Regimen

☐ ADR ☐ Literature Review Based

☐ Laboratory Data Findings ☐ Drug Interaction

9. Intervention Accepted

☐ Yes ☐ No

10 If 'No' to 8, Reason:

_____

_____

_____

10. Attending Physician's Name & Signature:

11. Results (Check all that apply):

Rx: ☐ Dispensed as written ☐ Clarified as Written

  ☐ Not Dispensed ☐ Changed and Dispensed

Patient: ☐ Counseled

  ☐ Written Information Given to Patient

Others: ☐ Improved compliance

☐ Increased therapeutic effectiveness

☐ Improved monitoring therapy

☐ Prevent toxicity/side effect

☐ Reduced cost by Rs._____

12. Follow up details of the Patient

_____

_____

_____

Intervention Made By   **Name:**

Designation:

Signature with Date:

Activity - 5
Patient Counselling

Patient counselling refers to the process of providing information, advice, and assistance to help patients use their medications appropriately. According to USP, medication counseling is an approach that focuses on enhancing the problem-solving skills of the patients to improve or maintain the quality of health and quality of life.

The information is usually given verbally but may be supplemented with written materials. During counseling, the pharmacist should assess the patient's understandings about his or her illness and treatment, and provide individualized advice and information which will assist the patient to take their medication most safely and effectively. To provide accurate advice and information, the pharmacist should be familiar with the pathophysiology and therapeutics of the patient's disease.

Good communication skills are needed to gain the patient's confidence and to help motivate the patient to adhere to the recommended regimen. Effective patient counselling aims to produce the following results 1) Better patient understanding of their illness and the role of medication in its treatment. 2) Improve medication adherence 3) more effective drug treatment 4) reduced incidence of adverse effects and unnecessary healthcare costs. 5) Improved quality of life for the patient. 6) Better coping strategies to deal with medication-related adverse effects. 7) Improved professional rapport between the patient and the pharmacist.

Communication skills for effective counselling Communication is the transfer of information meaningful to those involved. It is the process in which messages are generated and sent by one person and received and translated by another person. The communication process between health professionals and patients serves two primary functions. 1. It establishes an ongoing relationship between the professional and the patient. 2. It provides the exchange of information necessary to assess a patient's health condition, implement treatment of medical problems, and evaluate the effects of treatment on a patient's quality of life.

The healthcare professional must be able to - understand the illness experience of the patient - perceive each patient's experience as unique - good relationship with patients - build a therapeutic alliance with patients to meet mutually understood goals of therapy - develop self-awareness of personal effects on patients.

The Communication Process Uses 1) Verbal Communication Skills:- Like Language Tone Volume Speed 2) Non-Verbal Communication Skills:- Like Body Language Proximity Eye Contact Facial Expression.

Barriers to effective communication include Physical, Psychological, Administrative or Time Conflict. Such conflicts prevent effective communication from being established.

Steps during Patient Counselling:

The following are the steps to be followed during patient counselling

1) Preparing for the session.
2) Opening of the session.
3) Counselling content.
4) Closing the session.

Process Steps:

1) Establish caring relationships with patients as appropriate to the practice setting and stage in the patient's health care management. Introduce yourself as a pharmacist, explain the purpose and expected length of the sessions, and obtain the patient's agreement to participate. Determine the patient's primary spoken language.

2) Assess the patient's knowledge about his or her health problems and medications, physical and mental capability to use the medications appropriately and attitude toward the health problems and medications. Ask open-ended questions about each medication's purpose and what the patient expects, and ask the patient to describe or show how he or she will use the medication. They should also be asked to describe any problems, concerns, or uncertainties they are experiencing with their medications.

3) Provide information orally and use visual aids or demonstrations to fill patients' gaps in knowledge and understanding. Open the medication containers to show patients the colors, sizes, shapes, and markings on

oral solids. For oral liquids and injectables, show patients the dosage marks on measuring devices. Demonstrate the assembly and use of administration devices such as nasal and oral inhalers. As a supplement to face-to-face oral communication, provide written handouts to help the patient recall the information.

4) If a patient is experiencing problems with his or her medications, gather appropriate data and assess the problems. Then adjust the pharmacotherapeutic regimens according to protocols or notify the prescribers.

5) Verify patients' knowledge and understanding of medication use. Ask patients to describe or show how they will use their medications and identify their effects. Observe patients' medication-use capability and accuracy and attitudes toward following their pharmacotherapeutic regimens and monitoring plans.

Counselling Content:

1. The medication's trade name, generic name, common synonym, or other descriptive names (s) and, when appropriate, its therapeutic class and efficacy.

2. The medication's use and expected benefits and action. This may include whether the medication is intended to cure a disease, eliminate or reduce symptoms, arrest or slow the disease process, or prevent the disease or a symptom.

3. The medication expected onset of action and what to do if the action does not occur.

4. The medication's route, dosage form, dosage, and administration schedule (including duration of therapy).

5. Directions for preparing and using or administering the medication. This may include adaptation to fit patients' lifestyles or work environments.

6. Action to be taken in case of a missed dose.

7. Precautions to be observed during the medication's use or administration and the medication's potential risks concerning benefits. For injectable medications and administration devices, concern about latex allergy may be discussed.

8. Potential common and severe adverse effects that may occur, actions to prevent or minimize their occurrence, and actions to take if they occur, including notifying the prescriber, pharmacist, or other health care provider.

9. Techniques for self-monitoring of the pharmacotherapy.

10. Potential drug-drug (including nonprescription), drug-food, and drug-disease interactions or contraindications.

11. The medication's relationships to radiologic and laboratory procedures (e.g., timing of doses and potential interferences with the interpretation of results).

12. Prescription refill authorizations and the process for obtaining refills.

13. Instructions for 24-hour access to a pharmacist.

14. Proper storage of the medication.

15. Proper disposal of contaminated or discontinued medications and used administration devices.

16. Any other information unique to an individual patient or medication.

Patient Counseling Form Instructions

Patient counseling form refers to the process of providing vital information, advice, and assistance to help the patient with medication given by the pharmacist and ensuring to them properly. This also includes important information about the patient's illness lifestyle and lifestyle modifications the patient counseling form includes:

I.P.NO, Date, code no, age, sex, wt, past medical history, family medical history, social history, current illness, allergies(drug/food/other), current medication, counseling given on, patient prescription concerning disease and medication, patient compliance and evaluation, major side effects and management, counseling points

For example:

I.P NO: XXXX

Date: 1-1-2109

Code No: XYZ

Age: 45

Sex: F

Wt: 60

Past Medical History:

Family Medical History:

Social History:

Past Medication History:

Current Illness: Giant lipoma

Allergies (drug/food/other):

Current Medication: ertapenem, tramadol, ondansetron, Ultram, IVF-NS/RL,

Counseling Given on diet of the patient

Patient Perception concerning Disease and Medication (it is about the knowledge of the patient towards disease he/she have) example: patient has adequate knowledge of discomfort, fever, and regular intake of medication.

Patient compliance and evaluation (whether the patient recovery is poor (or) satisfactory (or) good should be mentioned): good

Major side effects and managements (if there are any major side effects observed in the medicated drugs, we should give counseling based upon the management of the side effect): there are no side effects

Counseling Points:

It includes the major counseling point which is given to the patient by the pharmacist that is:

Precautions, diet, and exercise, interactions (drug-drug, drug-food, drug-disease), storage, information on missed doses, any communication barriers.

For example:

a) Precaution: regular care should be taken at the site of the surgery
b) Diet and exercise: soft diet is prescribed to the patient
c) Interaction : no interactions
d) Storage: Store in a cool and dry place.
e) Information on missed doses: Take the drug as soon as possible before 2 hours; otherwise it is better to skip the drug.

Any Communication Barriers:

If there is any communication barrier it should be mentioned either yes or no.

If yes:

The communication barrier may include language, literacy, physical (sensory impairment), anxiety, age, time, non-co operation.

How was the Barrier Overcome?

Barrier should be overcome by improving physician-patient communication, information sharing, educational interventions, increasing focus on preventive research and exploring strategies to reduce adverse effects (for example dosing modifications).

Name of the Patient and Signature: Xyz

Name of the Pharmacist and Signature: Abc

Name of Faculty-Incharge: Efg

References:

Naresh Panigrahi, Assistant Professor, Gitam University. Patient Counselling. http://www.authorstream.com/Presentation/nareshph28-2412502-patient-counselling/

Patient Counselling

Patient Counselling Form

I.P.NO   Date: Code No:

Age: Sex: Wt:

Past Medical History:

Family Medical History:

Social History:

Past Medication History:

Current Illness:

Allergies (drug/food/other):

Current Medication:

Counseling is given on:

Patient perception concerning disease and medication:

Patient compliance and evaluation:

☐ Poor   ☐ Satisfactory   ☐ Good

Major side effects and management:

Counseling Points:

A. Precautions:

B. Diet and exercise:

C. Interactions (drug-drug, drug-food, drug-disease):

D. Storage

Information on Missed Doses:

Any Communication Barrier:

☐ Yes   ☐ No

If Yes:

☐ Language   ☐ Literacy

☐ Physical Sensory (impairment)   ☐ Anxiety

☐ Age   ☐ Time

☐ If no- cooperative

How was the barrier overcome?

Name of Patient: Sign:

Name of Pharmacist: Sign:

Name of Faculty-Incharge :   Sign:

Patient Counselling Form

Activity - 6
Drug Information Request and Documentation

The provision of drug information (DI) is among the fundamental professional responsibilities of all pharmacists. Drug information may be patient-specific, academic (for educational purposes), or population-based (to aid in the decision-making process for evaluating medication use for groups of patients). The goal of providing carefully evaluated, evidence-based recommendations to support specific medication-use practices is to enhance the quality of patient care, improve patient outcomes, and ensure the prudent use of resources [1].

Drug Information Activities

To be an effective provider of DI, the pharmacist must exercise excellent oral and written communication skills and be able to

1. Anticipate and evaluate the DI needs of patients and health care professionals.

2. Obtain appropriate and complete background information as described under the section Systematic Approach for Responding to Drug Information Requests.

3. Use a systematic approach to address DI needs by effectively searching, retrieving, and critically evaluating the literature (i.e., assessment of study design, statistics, bias, limitations, and applicability).

4. Appropriately synthesize, communicate, document, and apply pertinent information to the patient care situation. [2, 3].

Systematic Approach for Responding Drug Information Request

Systematic Approach for Responding to Drug Information Requests A systematic approach for responding to DI requests was first introduced by Watanabe, et al. in 1975 [4]. This approach has been modified and expanded over the years to ensure that all relevant information is considered before formulating a response [2, 5]. The importance of gathering pertinent patient data and understanding the context of a question before answering a DI request is described in the literature.14-16 Of note, a full systematic approach may not be practical for all requests, especially for urgent clinical needs in the direct patient care setting. Also, consideration should be given to the ethical and legal aspects of responding to DI requests, including patient privacy concerns [2]. A systematic approach may be outlined as follows [2,5].

1. *Identify the requestor.* To obtain complete information and develop a response with the appropriate perspective, consider the health literacy and professional background of the requestor.

2. *Define the true question and information need.* Identify the true question and information needed by asking probing questions of the requestor. For example, "Why is the question being asked?" and "Does the question pertain to a specific patient?" may help reveal important details of the true question.1 This kind of information helps in

optimizing the search process and assessing the appropriate time frame of response need.

3. *Obtain complete background information.* Obtain more complete background information, including examining the medical record for patient data, if applicable, to individualize the response to meet the requestor's need.

4. *Categorize the question.* Classify requests as patient-specific or academic and by type of question (e.g., product availability, adverse drug event, compatibility, compounding/formulation, dosage/administration, drug interaction [drug-drug, drug-disease, drug-laboratory], drug product identification, pharmacokinetics, therapeutic use/efficacy [FDA approved vs. unlabeled indications], safety in pregnancy/nursing toxicity/poisoning) to aid in tailoring the search strategy and selecting resources.

5. *Perform a systematic search.* Perform a systematic search of appropriate tertiary, secondary, and primary resources, including electronic resources, as necessary.

6. *Analyze the information.* Evaluate, interpret, and combine information from the resources used. Other information needs should be anticipated as a result of the information gathered.

7. *Disseminate the information.* Provide an oral or written response, or both, as needed by the requestor that specifically applies the information to a particular situation. The information, its urgency, and its purpose

may influence the method of response. Supporting documentation (e.g., primary literature) should be included when possible.

8. *Document.* Document the request, information resources used, the information found in each source, time spent on the response, and the response itself as appropriate for the request and the practice setting.

9. *Follow-up.* Perform a follow-up assessment to determine the utility of the information provided and whether the information resulted in changes in medication-use practices or patient outcomes.

Documentation and Quality Assessment

Numerous methods of documenting pharmacist interventions, including the provision of DI, have been described in the literature. Drug information centers are moving toward increased use of electronic documentation systems, which have helped to increase the depth and quantity of documentation, as well as provide increased efficiency and cost savings [6-10]. Also, an electronic system can promote a standardized and systematic approach and provides a readily retrievable archive that can be used to rapidly search previously answered questions [11, 12]. Documentation of DI services should incorporate elements identified through the systematic approach. The ASHP Guidelines on Documenting Pharmaceutical Care in Patient Medical Records states that "the professional actions of pharmacists that are intended to ensure safe and effective use of drugs and that may affect patient outcomes should be documented in the patient medical record [43]. Therefore, if the DI request is patient-specific, it is appropriate, but not always necessary, to document the request and response in the patient's medical record. Documentation is critical to appropriate patient care, highlights the value of pharmacist services, demonstrates accountability, provides a basis for quality assessment and performance improvement, and details an appropriate systematic approach in case a medico-legal dispute arises from a DI request. Consequently, even academic or population-based DI activities should be appropriately documented. Despite the importance of assessing the quality of drug-related information provided by pharmacists, there is currently no standardized method described in the literature. However, some DI centers have reported use of double-check systems before providing a response, random retrospective audits by a DI specialist or another individual, obtaining feedback from the requestor, and conducting an internal review by a committee as methods of quality assessment [15].

Resources

Pharmacists need to use appropriate and credible resources. The following factors should be considered when purchasing DI resources, including electronic subscriptions, for the pharmacy department or practice setting [1]:

1. Features of the resource (e.g., frequency of updates, qualifications, and affiliations of authors, year of publication, type of information, organization of material, method of delivery, cost).

2. Practice setting (e.g., type of facility and needs of health care professionals within that environment, state-specific regulatory requirements).

3. Accessibility of the resource (e.g., location of print resources, the number of users allowed by subscription).

Drug Information Request & Documentation Form Instructions

This form includes information about all drug-related information and documentation of the patient. The received data and received time should be mentioned by the pharmacist at the time of receiving queries by any professional people with status like physician, surgeon, resident, interns, pharmacist, nurse and others like PG's(specify). It includes details of the name of the enquirer, designation, phone no and unit. It should also include the query question.

For example:

Code no: xxxxxx

Received date: 11-11-19

Received time: 11:00 AM

Received by: the name of the pharmacist should be mentioned

Designation: trainee pharmacist

Phone no: 800**886

Unit: cubicals

Professional status: others (UG)

Query: the query should be related to the drug, the question should be based upon drug therapy, drug efficacy, type of drug, drug toxicity, contraindication of drug, adverse and side effects of drugs, etc.

For example:

What type of antibacterial drug is given to the patient in the case of a giant lipoma?

Information provided:

It should include information about the way of request by enquirer to the pharmacist i.e, mode of request through direct access(or) during ward rounds(or) telephone and others. It should also include the purpose of above inquiry for updated knowledge (or)better patient care (if it found then, the details should be mentioned), it should also include the duration of answering the above query by the pharmacist to the enquirer like immediately or within 2-4 hrs or 1-2 days or a day .if the duration of reply is late more than one day, the reason for delay should be mentioned.

For example:

Mode of request: direct access

Purpose of inquiry: updated knowledge

Answer given: immediately

Delay for answer(if any): none

Questions category:

It should include all the information about the questions based on a drug like drug therapy, indication, efficacy, pharmacokinetics /pharmacodynamics, dose/administration, pregnancy/lactation, poisoning, stability, identification, incompatibility

For example, the category of the query which is asked by the enquirer is about drug therapy.

Patient's details:

It includes the information about the patient like age, weight, sex, allergies, current medical problem and it also includes the detail information (or) function of renal/hepatic function, pregnancy/lactation (if it is yes we should mention) and other important investigation (abnormality of the lab investigation values and by vital signs value), drug therapy (for abnormal lab values drug therapy is given).

For example:

Age: 45

Weight: 60

Sex: F

Current medical problem: giant lipoma

Hepatic /renal function details: none

Pregnancy/lactation: no

Other important investigation: no abnormal values

Drug therapy: ertapenem, tramadol, ondansetron, Primera.

Other details:

Other details like if the patient is pregnant it should include the stage of trimester first (or) second (or) third and if breastfeeding, age of infant is mention.

Query response:

For the above-given query, the answer should be given

For example:

Ertapenem is given as an antibacterial drug in case of giant lipoma

Reference:

An answer resource to the query should mention its reference like a textbook (mention), journals (mention), Micromedex, IDS, website, other (specify)

For example the reference of the above query from Micromedex (drug information)

Mode of reply:

The mode of replying by the pharmacist can be 4 ways that can be in a written, verbal,(or)both and printed literature format

For example, the mode of reply by the pharmacist is verbal

Date of reply:

Reply date should be mentioned

Time of reply:

Reply time should be mentioned

Follow up (if any): we should follow up the patient for better information (or) knowledge of the case

Reporting date:

Reporting date should be mentioned

Reporting time:

Reporting time should be mentioned

Name of the pharmacist:

Name of the pharmacist should be mentioned

Signature of the pharmacist:

Signature of the pharmacist should be mentioned

References

1. Medication Therapy and Patient Care: Specific Practice Areas–Guidelines. https://www.ashp.org/-/media/assets/policy-guidelines/docs/guidelines/pharmacists-role-providing-drug-information.ashx

2. Malone PM, Kier KL, Stanovich JE. Drug information: a guide for pharmacists, 4th ed. New York: McGraw-Hill; 2012.

3. Bernknopf AC, Karpinski JP, McKeever AL, et al. Drug information: from education to practice. Pharmacotherapy. 2009; 29:331–46.

4. Watanabe AS, McCart G, Shimomura S, et al. Systematic approach to drug information requests. Am J Hosp Pharm. 1975; 32:1282–5.

5. Nathan JP. Drug information—the systematic approach: continuing education article. J Pharm Pract. 2013; 26:78–84.

6. Wisniewski CS, Pummer TL, Krenzelok EP. Documenting drug information questions using software for poison information documentation. Am J Health-Syst Pharm. 2009; 66:1039–43.

7. Simonian AI. Documenting pharmacist interventions on an intranet. Am J Health-Syst Pharm. 2003; 60:151–5.

8. Nurgat ZA, Al-Jazairi AS, Abu-Shraie N, et al. Documenting clinical pharmacist intervention before and after the introduction of a web-based tool. Int J Clin Pharm. 2011; 33:200–7.

9. Abe AM. Implementation of a cloud computing system for documenting drug information consultation requests and responses. Poster presented at ASHP Midyear Clinical Meeting; December 2012; Las Vegas, NV.

10. Brown JN. Cost savings associated with a dedicated drug information service in an academic medical center. Hosp Pharm. 2011; 46:680–4.

11. Erbele SM, Heck AM, Blankenship CS. Survey of computerized documentation system use in drug information centers. Am J Health-Syst Pharm. 2001; 58:695–7.

12. Cheng SC. Computerized tools that standardize the systematic approach to researching drug information requests. Poster presented at ASHP Midyear Clinical Meeting; December 2006; Anaheim, CA.

13. American Society of Health-System Pharmacists. ASHP guidelines on documenting pharmaceutical care inpatient medical records. Am J Health-Syst Pharm. 2003; 60:705–7.

14. Rosenberg JM, Koumis T, Nathan JP, et al. Current status of pharmacist-operated drug information centers in the United States. Am J Health-Syst Pharm. 2004; 61:2023–32.

Drug Information Request and Documentation Form

Code No: **Received date:**

Received time: **Received By:**

Name of the Enquirer:

Name of Enquirer :

Designation :

Phone No :

Unit :

Professional status :

☐ Physician   ☐ Surgeon   ☐ Resident

☐ Interns   ☐ Pharmacist   ☐ Nurse

☐ Others PG's (specify)

Query:

Information Provided :

Mode of request:

☐ Direct Access   ☐ During Ward rounds

☐ Telephone   ☐ Others

Purpose of inquiry:

☐ Updated Knowledge

☐ Better Patient Care(if yes give details below)

☐ Others

Answer given

☐ Immediately ☐ Within 2-4 hrs

☐ Within 1-2 days ☐ Within a day

Delay for answer (if any)

Questions category :

☐ Drug therapy ☐ Pregnancy/ lactation

☐ Indications ☐ Poisoning

☐ Efficacy ☐ Stability

☐ Pharmacokinetics/ Pharmacodynamics

☐ Identification

☐ Dose / administration ☐ Incompatibility

Patients details:

Age(yrs): weight(kgs): Sex : M/F

Allergies :

Current medical problem:

Hepatic / renal function details:

Pregnancy/ lactation: Y/N (If yes give details)

Other important investigations:

Drug therapy:

Other Details:

If Pregnant :　　☐ First Trimester　　☐ Second Trimester

　　　　　　　　☐ Third Trimester

　　　　　　　　☐ If Breast Feeding, Age of Infant:

Query response:

References :

Textbook (mention):

Journals (mention):

Micromedex :

IDIS :

Website:

Others (specify):

Mode of reply :

☐ Written　　☐ Verbal　　☐ Both　　☐ Printed literature

Date of Reply:

Time of Reply:

Follow up (if any):

Reporting Date: **Reporting Time:**

Name of pharmacists: **Signature of the pharmacists :**

Activity – 7
Patient History Details

The content of the history required in primary care consultations is very variable and will depend on the presenting symptoms, patient concerns and past medical, psychological and social history. However, the general framework for history taking is as follows [1].

➢ Presenting complaint.

➢ History of presenting complaint, including investigations, treatment and referrals already arranged and provided.

➢ Past medical history: significant past diseases/illnesses; surgery, including complications; trauma.

➢ Medication history: now and past, prescribed and over-the-counter medicines, allergies.

➢ Family history: especially parents, siblings, and children.

➢ Social history: smoking, alcohol, recreational drugs, accommodation and living arrangements, marital status, baseline functioning, occupation, pets, and hobbies.

➢ Systems review cardiovascular system, respiratory system, gastrointestinal system, nervous system, musculoskeletal system, genitourinary system.

There are several consultation models that are useful to frame (and remember) your questions. Medical schools in the UK often use the Calgary-Cambridge model [2, 3].

Review of systems

Whatever system a specific condition may seem restricted to, all the other systems are usually reviewed in a comprehensive history. The review of systems often includes all the main systems in the body that may provide an opportunity to mention symptoms or concerns that the individual may have failed to mention in history. Health care professionals may structure the review of systems as follows:

➢ Cardiovascular system (chest pain, dyspnea, ankle swelling, palpitations) are the most important symptoms and you can ask for a brief description for each of the positive symptoms.

➢ Respiratory system (cough, hemoptysis, epistaxis, wheezing, pain localized to the chest that might increase with inspiration or expiration).

➢ Gastrointestinal system (change in weight, flatulence, and heartburn, dysphagia, odynophagia, hematemesis, melena, hematochezia, abdominal pain, vomiting, bowel habit).

➢ Genitourinary system (frequency in urination, pain with micturition (dysuria), urine color, any urethral discharge, altered bladder control like urgency in urination or incontinence, menstruation and sexual activity).

➢ Nervous system (Headache, loss of consciousness, dizziness and vertigo, speech and related functions like reading and writing skills and memory).

> Cranial nerve symptoms (Vision (amaurosis), diplopia, facial numbness, deafness, oropharyngeal dysphagia, limb motor or sensory symptoms and loss of coordination).

> Endocrine system (weight loss, polydipsia, polyuria, increased appetite (polyphagia) and irritability).

> Musculoskeletal system (any bone or joint pain accompanied by joint swelling or tenderness, aggravating and relieving factors for the pain and any positive family history for joint disease).

> Skin (any skin rash, recent change in cosmetics and the use of sunscreen creams when exposed to sun).

Patient History Form Instructions

Patient history form includes the demographic details of the patient like date, patient ID, past medication history, past medical history, social history, this form is helpful for better patient care in case of past medical history

For example:
Name: Xyz
Age: 60
Patient Id: 12****
Date: 1-1-2019
Past medication history:

The past medication history includes the name of the medication, direction of use, starting date, stopping date, and the purpose of the medication.

For example:

Name	Directions	Start Date	Stop Date	Purpose
Furosemide	p/o	1-12-2015	1-1-2018	Antihypersensitive

Past medical history:

If there are any past medical history it should be mentioned like known hypertension, known DM, known kidney problem, known liver problem, frequent urination, difficulty with urination, frequent urination at night, nausea(or) vomiting, constipation (or) diarrhea, bloody(or) black bowel, frequent heartburn, cough with sputum, fainting in the past, sores on leg(or) foot, leg pain (or) swelling, anaemia, thyroid problem, muscle cramp (or)pain, dizziness, change in appetite, rash, walking or balance problem, TB, other

For example:
The past medical history of a particular patient is type 2 DM.

Social history:

Social history includes the lifestyle habitat of the patient like consumption of alcohol, nicotine use, caffeine use, the diet of the patient

For example, Consumption of alcohol should be avoided in case of type 2 diabetes.

Clinical pharmacist:

The name of the clinical pharmacist should be mentioned.

Signature:

A clinical pharmacist signature should be done.

References

1. The GP consultation in Practice, Royal College of General Practitioners, 2014.

2. Liu C, Scott KM, Lim RL, et al. EQ Clinic: a platform for learning communication skills in clinical consultations. Med Educ Online. 2013. Doi: 10.3402/meo.v21.1801.e collection 2016.

3. Wild D, Nawaz H, Ullah S et al: Teaching residents to put patients first: creation and evaluation of a comprehensive curriculum in patient-centered communication. BMC Med Educ. 2018. 1918 (1):266. DOI: 10.1186/s12909-018-1371-3.

Patient History Form

Patient Id: Date:

Name: Age: Sex:

Past Medication History:

Drug	Directions	Start Date	Stop Date	Purpose

Past Medical History:

☐ Known Hypertension_____

☐ Known DM_____

☐ Known kidney problem_____

☐ Known liver problem_____

☐ Frequent urination_____

☐ Difficulty with urination

☐ Frequent urination at night_____

☐ Nausea or vomiting

☐ Constipation or diarrhea

☐ Bloody or black bowel

☐ Frequent heartburn

☐ Cough with sputum

☐ Fainting in the past

☐ Sores on leg or foot

☐ Leg pain or swelling

☐ Anaemia

☐ Thyroid problem_____

☐ Muscle cramps or pain

☐ Dizziness

☐ Change in appetite

☐ Rash

☐ Walking or balance problem

☐ TB

☐ Others_____

Family history of any of the above problems_____

Social History:

Nicotine use: _____   Caffeine: _____

Alcohol consumption: _____   Diet: _____

Clinical Pharmacist: **Signature:**

Patient History Form

Activity – 8
Patient Referral Details

For hospitals and health care providers, and effective patient referral system is an integral way of ensuring that patients receive optimal care at the right time and at the appropriate level, as well as cementing professional relationships throughout the health care community.

Unfortunately, like a chain, a referral system is only as strong as its weakest link. Often referral systems contain many weak links – like using antiquated fax machines to process referrals – that compromise the system and make it unnecessarily difficult for providers and patients to navigate.

With a focused effort and the right resources, it is possible to create a modern, efficient and timely referral process that enhances office practices and increases patient satisfaction and referral compliance. The route a highly regarded health care system took to overhaul its referral processes offers several lessons that other organizations can heed when confronting a similar need.

The health care system's issues stemmed from a lack of a standardized, enterprise-wide process for handling referrals. The organization relied on a decentralized approach, with each of its individual clinics responsible for its own referral traffic. The health care system needed a better portal to support referring to providers' needs, which would significantly improve the handling of incoming patient referrals.

5 Steps to Referral Resolution

Sequentially, here is how the health care system successfully improved its referral system; these same steps can be used by any other health care organization interested in achieving similar gains.

1. Identify current and desired state – Before determining what is needed for a desired future state, you need to first review and assess the current state. The health care system created a design team to determine how the current referral system worked or did not, where and how referral process changes needed to be made, who would be handling them, and what next steps were needed.

2. Chart your desired future course – Over more than 20 meetings, the design team discussed, charted and determined proper referral handling scenarios. These scenarios ranged from the seemingly simple, such as a referral to an orthopedic specialist for a fractured limb, to the more complicated, such as the referral of a patient with multiple complex chronic illnesses. All possible types of referrals needed to be accounted for, to ensure that they would be handled appropriately and consistently.

3. Ensure provider portal functionality – The health care system owned a provider portal module, and was not using it to its fullest capability. Instead, the health care system relied on a fax-based referral system which required staffers to manually re-enter referrals into a separate database. This process was time-consuming, error-prone and a major bottleneck. Instead, the provider portal module offered a web-based application for connecting the health care system to local providers and providing them with secure access to the system's EMR. The design team ensured that this module would be regularly used in the future.

4. Create a new standardized process – Over several months, the design team created a new, standardized process for incoming referrals and leveraged the built-in referral benefits of the provider portal module. Critically, this included plans for internally training more than 150 staff members on this new process, as well as internal and external marketing of the process. The latter was vital to overcoming referring providers' previous impressions of the health care system mishandling patient referrals.

5. Gain top-level support – Owing to the vital clinical and business importance of having a high-performing referral process, the health care system's revamp efforts were continuously monitored and championed by C-suite executives (including the CMO, CIO, and CFO) as well as the vice president of nursing, vice president of ambulatory services, and many other executives. The buy-in and support of these leaders was important for providing the design team with the resources needed to successfully create this revamp.

Institutional Hospital Activities for Pharmacy Practice

Patient Referral Documentation Form

Date: **Time:**

Patient referred by (Dept.):

☐ Medicine ☐ E.N.T

☐ Pediatric ☐ Orthopedic

☐ Psychiatry ☐ Ophthalmology

☐ Pulmonology ☐ O.B.G

☐ Skin ☐ Surgery

Name of the Referrer:

Status of Referrer: ☐ Physician ☐ Surgeon

☐ Resident ☐ PG's

☐ Interns ☐ Nurse

☐ Pharmacist ☐ Others

Patient details:

Types of patients: ☐ Inpatient ☐ Outpatient

IP/OP No.:

Age: **Sex: M / F** **Weight:**

Allergies:

Current Medical Problems:

Relevant Drug Therapy:

Referred for: ☐ ADR detection and management
                     ☐ Patient Counseling
                     ☐ Management of Poisoning and toxicity
                     ☐ Selection of the Drug therapy
                     ☐ Others (please specify)

Patient background information collected? ☐ Yes ☐ No

Reason(s) for referral:

Opinion given:

Opinion was given within a specified time? ☐ Yes ☐ No

Opinions accepted: ☐ Yes ☐ No

Date:

Reference Consulted:

Follow up:

Attending Pharmacist   **Signature:**

Staff in-charge: **Signature:**

Patient Referral Documentation Form

Activity - 9
Patient Counselling Quality Assurance

Giving sufficient advice to patients about their medications has become a shared responsibility among healthcare providers, including community pharmacists [1,2,3]. If community pharmacists play a more active role in patient counseling, it can lead to improvements in the wellbeing and safety of patients and ultimately improve the quality of care while reducing patient nonadherence, treatment failure, and wasted health resources [4, 5, 6]. Although patient counseling has received a high priority in professional strategies and health policy goals [7, 8, 9], the implementation of new medication counseling behaviors in real-life practice has proved to be challenging [10, 11, 12].

To develop and improve the quality of counseling practices, specific quality assurance instruments are needed in community pharmacies. Pharmacists are aware of the principles of quality in the manufacturing and compounding of medicines, but it seems to be difficult to extend these principles to customer service, especially to professional services such as patient counseling. Moreover, most of the quality management systems in community pharmacies lack concrete indicators for such services. In healthcare or ganizations, quality management systems, such as total quality management (TQM) and Balanced Scorecard, are used to improve the quality of the services [13]. Even though these systems are seldom applied in community pharmacies, professional organizations have published principles of practice to promote the implementation of professional services and improves the quality of counseling [14].

An inventory of existing patient counseling–specific instruments reveals that they focus primarily on assessing patient satisfaction [15, 16, 17]. They are limited to pharmacist-patient interaction and do not help in assessing other factors influencing performance (eg, premises, skills and knowledge of the personnel, dispensing process).

Quality assurance is achievable through ongoing evaluation of patient care which would assure the hospital that all that was done for the patient.

Quality assurance helps patients by improving quality of care in the following ways

1. Assess the competence of medical staff, serve as an impetus to keep up to date and prevent future mistakes.
2. Bring to notice of hospital administration the deficiencies and in correcting the causative factors.
3. Help to exercise a regulatory function.
4. Restricting undesirable procedures.

Quality assurance is a never-ending process of creative destruction, with rapid advances in science and technology and medical knowledge continuous updating is essential. The emphasis is on establishing professional excellence patient satisfaction at a reasonable cost. Quality is not proportionate to the use of sophisticated technology or to be an expense incurred. The technical imperative should not insist on prolonging life at any loss with no consideration to quality of life.

Patient Counselling Quality Assurance form Instructions

Patient counseling is a professional service provided by the community pharmacist. To maintain the quality of these services, specific quality assurance instruments are needed and in this form, it includes about the demographic details of the patient and diagnosis of the patient, it also contains some questions like:

1. Whether counselling steps were followed? (Mention yes (or) no)

2. Whether counselling points were covered during counselling session? (Mention yes (or) no)

3. Whether counselling aids were provided? (Mention yes (or) no)

4. Whether counselling material were provided? (Mention yes (or) no)

5. Whether the barriers were rightly overcome? (Mention yes (or) no)

6. Was the time taken for counselling appropriate? (Mention yes (or) no)

7. Was the understanding of the patient ascertained? (Mention yes (or) no)

8. Was the process of the patient counselling properly documented? (Mention yes (or) no)

Grade:

Grade should be given according to the points like

An (excellent), B (good), c (can improve), D (should improve)

Note: A=8 points, B=5-6 points, c=3-4 points, &D=<3 points

Remarks:

If any remarks found they should be mentioned under this column.

Name of the counseling pharmacist/student:

Name of the counseling pharmacist/student should be mentioned

Name of the preceptor:

Name of the preceptor should be mentioned

Date/signature:

Signature of the preceptor with date should be mentioned

References

1. Bissel P, Ward PR, Noyce PR. Appropriateness measurement: application to advice-giving in community pharmacies. Soc Sci Med 2000;51:343-59.
2. Nichol MB, Michael LW. Critical analysis of the content and enforcement of mandatory consultation and patient profile laws. Ann Pharmacother 1992;26:1149-55.
3. Smith FJ, Salkind MR, Jolly BC. Community pharmacy: a method of assessing the quality of care. Soc Sci Med 1990; 31:603-7.
4. Cox K, Stevenson F, Britten N, Dundar T. A systematic review of communication between patients and health care professionals about medicine-taking and prescribing. www.concordance.org (accessed 2005 May 27).

5. Roter DL, Hall JA, Merisca R, Nordstrom B, Cretin D, Svarstad B. Effectiveness of interventions to improve patient compliance: a meta-analysis. Med Care 1998;36: 1138-83.

6. Roughead L, Semple S, Vitry A. The value of pharmacist professional services in the community setting. A systematic review of the literature 1990–2002. Quality Use of Medicines and Pharmacy Research Centre. University of South Australia, Adelaide, 2002.

7. International Pharmaceutical Federation (FIP). FIP standards for quality of pharmacy services. www.fip.org (accessed 2005 May 27).

8. Guidelines for professional community pharmacy. Association of Finnish Pharmacies, Helsinki, Finland: 1997.

9. National Health Service (NHS). Pharmacy in the future. Implementing the NHS Plan. A programme for pharmacy in the National Health Service. www.doh.gov.uk/ pharmacy future (accessed 2004 Jan 15).

10. Morris LA, Tabak ER, Gondek K. Counseling patients about prescribed medication: 12-year trends. Med Care 1997;35:996-1007.

11. Erickson SR, Kirking DM, Sandusky M. Michigan Medicaid recipients' perception of medication counseling as required by OBRA '90. J Am Pharm Assoc 1998;38:333-8.

12. Goodburn E, Mattosinho S, Mongi P, Waterston T. Management of childhood diarrhoea by pharmacists and parents: is Britain lagging behind the Third World? BMJ 1991;302:440-3.

13. Zelman WN, Pink GH, Matthias CB. Use of the Balanced Scorecard in health care. J Health Care Finance 2003;29:1-16.

14. American Pharmaceutical Association (APhA). Principles of practice for pharmaceutical care. www.aphanet.org (accessed 2005 Jan 20).

15. Ward PD, Bissel P, Noyce PR. Criteria for assessing the appropriateness of patient counseling in community pharmacies. Ann Pharmacother 2000;34:170-5. DOI 10.1345/aph.19135

16. De Young MH. Reflections on guidelines and theories for pharmacist-patient interactions. J Pharm Teaching 1996;5:59-80.

17. Larson L, Rovers J, MacKeigan L. Patient satisfaction with pharmaceutical care: update of a validated instrument. J Am Pharm Assoc 2002;42: 44-50.

Patient Counseling Quality Assurance Form

Name: Age: Sex:

Patient Id: Date:

Diagnosis:

1. Whether counseling steps were followed?

   ☐ Yes   ☐ No

2. Whether counseling points were covered during counseling sessions?

   ☐ Yes   ☐ No

3. Whether counseling aids were provided?

   ☐ Yes   ☐ No

4. Whether counseling materials were provided?

   ☐ Yes   ☐ No

5. Whether the barriers were rightly overcome?

   ☐ Yes   ☐ No

6. Was the time taken for counseling appropriate?

   ☐ Yes   ☐ No

Patient Counselling Quality Assurance Form

7. Was the understanding of the patient ascertained?

   ☐ Yes   ☐ No

8. Was the process of patient counseling properly documented?

   ☐ Yes   ☐ No

GRADE: A (Excellent), B (Good), C (Can improve), D (Should improve)

NOTE: A=8 points, B=5-6 points, C=3-4 points & D=<3 points

REMARKS:

Name of the counseling Pharmacist/Student:

Name of the Preceptor: **Date/Signature**

Activity – 10
Prescription Details

A prescription, often abbreviated R or Rx, is a health-care program implemented by a physician or other qualified health care practitioner in the form of instructions that govern the plan of care for an individual patient. The term often refers to a health care provider's written authorization for a patient to purchase a prescription drug from a pharmacist [1].

Prescription to be accepted as a legal medical prescription, it needs to be filed by a qualified dentist, herbalist, advance practice nurse, pharmacist, physician, veterinarian, etc., for whom the medication prescribed is within their scope of practice to prescribe such treatments. This is regardless of whether the prescription includes controlled substances or over-the-counter treatments.

Prescriptions may be entered into an electronic medical record system and transmitted electronically to a pharmacy. Alternatively, a prescription may be handwritten on preprinted prescription forms that have been assembled into pads or printed onto similar forms using a computer printer or even on plain paper according to the circumstance. In some cases, a prescription may be transmitted from the physician to the pharmacist orally by telephone; this practice may increase the risk of medical error. The content of a prescription includes the name and address of the prescribing provider and any other legal requirement such as a registration number.

Each prescription is dated and some jurisdictions may place a time limit on the prescription. In the past, prescriptions contained instructions for the pharmacist to use for compounding the pharmaceutical product but most prescriptions now specify pharmaceutical products that were manufactured and require little or no preparation by the pharmacist. Prescriptions also contain directions for the patient to follow when taking the drug. These directions are printed on the label of the pharmaceutical product [8].

The word "prescription", from "pre-" ("before") and "script" ("writing, written"), refers to the fact that the prescription is an order that must be written down before a compound drug can be prepared. Those within the industry will often call prescriptions simply "scripts".

Many brand name drugs have cheaper generic drug substitutes that are therapeutically and biochemically equivalent. Prescriptions will also contain instructions on whether the prescriber will allow the pharmacist to substitute a generic version of the drug. This instruction is communicated in a number of ways.

The protocol is for the prescriber to handwrite one of the following phrases: "dispense as written", "DAW", "brand necessary", "do not substitute", "no substitution", "medically necessary", "do not interchange" [3].

Some prescribers further inform the patient and pharmacist by providing the indication for the medication; i.e. what is being treated.

This assists the pharmacist in checking for errors as many common medications can be used for multiple medical conditions.

Some prescriptions will specify whether and how many "repeats" or "refills" are allowed; that is whether the patient may obtain more of the same medication without getting a new prescription from the medical practitioner. Regulations may restrict some types of drugs from being refilled.

References

1. Belknap, SM; Moore, H.; Lanzotti, SA; Yarnold, PR; Getz, M.; Deitrick, DL; Peterson, A.; Akeson, J.; Maurer, T.; Soltysik, RC; Storm, GA; Brooks, I. (2008). "Application of Software Design Principles and Debugging Methods to an Analgesia Prescription Reduces Risk of Severe Injury from Medical Use of Opioids". *Clinical Pharmacology & Therapeutics*. **84** (3): 385–392.

2. "Guide to Good Prescribing - A Practical Manual: Part 3: Treating your patients: Chapter 9. STEP 4: Write a prescription". *apps.who.int*. Retrieved 26 March 2018.

3. "State Laws or Statutes Governing Generic Substitution by Pharmacists". : Epilepsy.com/Professionals. 2007-04-25.

Prescription Details

Prescription Slip

Department of Pharmacy Practice

Name: **Date:**

Age: **Sex: (M/F)**   **Ht:**   **Wt:**   **BMI:**

BP: **HR:**   **PR:**   **Temp:**

Diagnosis:

RX:

Trainee pharmacist:

Signature: **Date:**

Prescription Slip